Crislaine Ferrari

Use of natural coagulants any treatment of effluent

Crislaine Ferrari

Use of natural coagulants any treatment of effluent

LAP LAMBERT Academic Publishing

Imprint

Any brand names and product names mentioned in this book are subject to trademark, brand or patent protection and are trademarks or registered trademarks of their respective holders. The use of brand names, product names, common names, trade names, product descriptions etc. even without a particular marking in this work is in no way to be construed to mean that such names may be regarded as unrestricted in respect of trademark and brand protection legislation and could thus be used by anyone.

Cover image: www.ingimage.com

Publisher:
LAP LAMBERT Academic Publishing
is a trademark of
International Book Market Service Ltd., member of OmniScriptum Publishing Group
17 Meldrum Street, Beau Bassin 71504, Mauritius

Printed at: see last page
ISBN: 978-613-9-95205-2

Copyright © Crislaine Ferrari
Copyright © 2018 International Book Market Service Ltd., member of OmniScriptum Publishing Group

Literature Review

Uso de coagulantes naturais no tratamento de efluente da indústria de alimentos em substituição ao cloreto férrico

Use of natural coagulants in the treatment of food industry effluent replacing ferric chloride

índice

Resumo 3
Abstract 5
Introduction 7
Literature review 11
    Effluents of food industry 11
    Treatments of effluents 12
    Coagulation 15
    Chemical coagulants 17
    Natural coagulants 19
    Chemicals coagulants x Natural coagulants 27
Conclusion 31
References 33

Resumo

Com o aumento da população mundial a produção de alimentos também tem aumentado para suprir essas necessidades. Como consequência desses fatores temos o aumento da utilização dos recursos naturais, entre eles pode-se destacar o hídrico que após ser utilizado durante todo o processo acaba gerando efluentes com alta carga poluente que se não forem devidamente tratados podem acarretar em grandes problemas ambientais. Para minimizar esse problema faz-se o tratamento destes efluentes e um dos mecanismos utilizados é a coagulação. Os agentes coagulantes podem ser de origem química como alumínio ou ferro, ou de origem natural, também chamados de polieletrólitos, como moringa, quitosana e tanino. As vantagens da utilização dos polieletrólitos são inúmeras, pois podem ser utilizadas em tratamento de efluentes de vários segmentos da indústria, tratamento de esgotos domésticos e até mesmo para tratamento de água potável, além de apresentarem menor custo, maior eficiência e, menor volume de logo gerado comparado com os coagulantes de origem química, além de menor toxicidade podendo ser descartado mais facilmente e, até mesmo, ser utilizado na agricultura como fertilizante.

Palavras-chaves adicionais: cloreto férrico; moringa; polieletólitos; quitosana; tanino.

Abstract

With increasing of world population food production has also increased to supply these needs. As a result of these factors have increased the use of natural resources, among them can be highlight that water after being used throughout the process eventually generates effluents with high pollutant load which if not properly treated can result in major environmental problems. To minimize this problem is used the treatment of these effluents and one of the mechanisms used is coagulation. The coagulants agents can be of chemical origin such as aluminum or iron, or natural origin, also called polyelectrolytes, such as moringa, chitosan and tannin. The advantages of using polyelectrolytes are numerous, they can be used in the effluents treatment from various industries, wastewater treatment and even drinking water treatment, in addition to their lower cost, higher efficiency, the lower volume of the sludge generated compared with the chemical coagulant, also its lower toxicity enables easier discharge and may even be used in agriculture as fertilizer.

Additional keywords: ferric chloride; moringa; polyelectrolyte; chitosan; tannin.

Introduction

The environmental concern is increasingly widespread. In this scenario world population has been increasing, which demand for processed and industrialized products, whose production requires large amounts of natural resources and water is the main resource used. This resource, after use throughout processing and refining, returns to the receiving bodies, and can cause water contamination if were not properly treated, damaging the entire surrounding environment.

Among manufacturing and processing sectors, agribusiness has been shown as an important segment of the global economic activity, especially animal processing (Fernandes, 2007). The cattle, pig and poultry processing can be mentioned as the main activities in Brazil. According to indicators of Instituto Brasileiro de Geografia e Estatística (IBGE, 2013) the results obtained in the study showed increasing slaughter rate, mainly cattle, pigs and poultry, since the creation of the survey in 1997. Consequently, amount of water used have been causing increased effluent volume at the end of the process.

In the frigorific industries, large amounts of water are used during the whole process such as in cleaning equipment, room/sites within the industry, animal receiving local and washing carcasses. The amount of available water for slaughter is 850 liters per slaughtered pig, defined by the Ministério de Agricultura, Pecuária e Abastecimento - MAPA in Ordinance No. 711, of November 1, 1995 (Brasil, 1995).

Therefore, due to the large volume of effluent generated its contamination and with the environmental regulations, this waste must be necessarily treated before being intended to the receiving body in order to have no sanitary and environmental damages. One of the processes used for the treatment of effluent is coagulation, being used for color, turbidity and natural organic matter removal (Budd et al., 2004, Da Silva et al., 2004, Hassemer et al., 2002, Huang et al., 2009, Leiknes, 2009).

Aluminum sulfate, ferric chloride, aluminum chloride, ferric sulfate and chlorinated ferrous sulfate are among the most commonly inorganic anticoagulants used, and others of organic origin, also known as polyelectrolytes, such as the extract of moringa seed, tannin and chitosan (Bongiovani et al., 2010, Matos et al., 2007, Pavanelli & Bernardo, 2002).

Coral et al. (2009) reported that prolonged use of aluminum sulfate has been questioned due to the presence of residual aluminum in the treated water and the sludge generated, often in high concentrations, at the end of the process which hamper its provision in the soil due to contamination and accumulation of this metal.

Natural coagulants has showing advantages over chemical coagulants, specifically regarding biodegradability, low toxicity and low residual sludge production rate (Moraes, 2004).

Inorganic coagulants are increasingly being replaced by products of organic origin due to excessive concern with natural resources and the strict standards determined by environmental legislation. This study is justified by searching clean technologies for wastewater treatment, thus allowing food production in a sustainable form with environmentally responsible. The aim of this study is a literature review of the advantages and disadvantages of using each coagulant, and the pH parameters and concentration that become more efficient as well.

Literature review

Effluents of food industry

The effluents of the food industry are characterized by high concentrations of oil and grease, sulfates, nitrates and phosphates and as a result, have high COD and BOD. Suspended solids, floating material, low biodegradability and high temperatures are still present. For containing blood, various tissues and fats are highly putrescible, entering in decomposition immediately after its formation, and releasing as unpleasant odors as their appearance, causing serious problems in anaerobic and aerobic processes (Jung et al., 2002, Naime & Garcia, 2005, Vidal, 2000).

According to Scarassati et al. (2003) the effluent of pigs cold-slaughterhouse is characterized by increased COD level (800 to 32.000 mg.L^{-1}); large oils and greases presence; floating materials (fat); high concentration of sedimentary and suspended solids; high concentration of organic nitrogen; coarse solid presence; pathogenic microorganisms presence.

The Conselho Nacional do Meio Ambiente – CONAMA, on its Resolution n° 430 of May, 13, 2011, complements the Resolution n° 357 of March, 17, 2005, providing that direct discharge of effluents from polluting sources into receiving bodies, can be only made after proper treatment (Brasil, 2011).

Treatments of effluents

The effluents treatment can be performed by chemical-physical and/or biological processes. Treatment by physical processes results in oil, grease and coarse material removal, being carried out with strainers and fat boxes. The biological treatment has as main function the organic matter removal, being used in effluents with large amounts of organic material easily biodegradable (Naime & Garcia, 2005).

According to Companhia Ambiental do Estado de São Paulo (CETESB, 2008) the treatment of cold- slaughterhouse effluents usually involves the following steps:

- Primary treatment: for the coarse removal of sedimentary, suspended and floating solids using equipment such as fat boxes and strainers.
- Equalization: this step is performed in a tank that allows absorb significate variation of flow and pollution load.
- Secondary treatment: in this stage, colloidal, dissolved and emulsified solids are removed, through biological action due to effluent biodegradable characteristics from previous treatments.
- Tertiary treatment: promotes supplemental solid, nutrients and pathogenic microorganisms.

Machado (2005) says that the tertiary treatment is necessary to, after the treatment, the effluent can be used in direct or indirectly reuse in the industrial plant. You can also use this treatment in the case of receiving bodies do not tolerate the loads of pollutants secondary treatment, even though it complies with the law.

It can mention as examples of tertiary treatments:

- Cloraction – this method uses chlorine, which penetrates in microorganisms cells and react with their enzymes destroying them, and also acting in iron and manganese oxidation, hydrogen sulphide removal, odor, color and flavor control and algae removal (Jordão & Arruda, 1995, Macêdo, 2001).
- Ultraviolet radiation – electromagnetic spectrum region which has UV radiation is particularly indicated to microorganisms inactivation, reaching its highest efficiency at 260 nm wavelength (Skoog, 1994).
- Reverse Osmosis – According to Schneider (2001) is used to desalinate sea, brackish and surface waters. The pressure applied must overcome the osmotic pressure of the solution to separate the salts of water.
- Ozonization – the ozonization involves two reactive mechanisms. Ozone direct attack and the attack through OH radicals formed in the ozone decomposition (Gogate & Pandit, 2004).
- Electrolytic treatment – consisted of electric energy conversion into chemical through electrolytic cell, where a

continuing current from external source induces oxide reduction reaction non-spontaneous (Hemkemeier et al., 2009).

- Coagulation/Flocculation – coagulation corresponds to dispersion colloidal destabilization obtained by reduction of repulsion forces between negative charge cells through addiction of chemical or natural coagulants, in this case occurs a fast mixture of coagulant with the environment and later at this stage, occurs the flake formation, named flocculation (Pavanelli, 2001, Silva, 2005).

Coagulation

Coagulation is consisted of physical actions and chemical reactions set, lasting a few seconds, between the coagulant, usually an aluminum salt or iron, water and its impurities, and it is presented in three phases: (i) forming salt hydrolyzed species when dispersed in water, (ii) destabilizing colloidal and suspended particles dispersed and in liquid mass and (iii) aggregation of these particles to form the flakes (Santos et al., 2007).

According to Di Bernardo and Dantas (2005) coagulation results from individual or combined action of four different mechanisms: compression of the diffuse layer, adsorption and neutralization, scan and adsorption and bridging:

- Compression of diffuse layer - when introducing simple salts in a colloidal, the charge density in the diffuse layer increases and sphere of particle influence decreases, causing coagulation by compressing the diffuse layer.
- Adsorption and neutralization – this mechanism is very important when applying direct filters technologies because there is no need to flakes production for further sedimentation, but destabilized particles that will be retained in filter granular median.
- Scan – it is widely used in treatment water stations where there is flocculation and sedimentation preceding filtration, i.e. in the cases where aluminum sulfate is applied.
- Adsorption and bridging - this mechanism involves the polymer use of great larger molecular chains, which serve as bridge between the surface which they are and other particles.

Schoenhals et al. (2006) defined coagulation/flocculation as the provider of the destabilization of colloidal and finely divided particles, thus forming larger and denser flakes, being possible the separation.

Santos (2004) reports that the coagulation diagrams are useful tools for predicting the chemical conditions where coagulation occurs and therefore essential to plan, analyze and interpret studies in laboratory scale or pilot scale, then contributing to the definition of coagulant dosage and pH conditions for reducing turbidity and color and other parameters of interest.

Chemical coagulants

Ferric choride

Iron salts are widely used as coagulants for water treatment. React in order to neutralize the negative charges of colloids and provide the formation of insoluble iron hydroxides. Due to the low solubility of the formed ferric hydroxides, they can act on a wide pH range. In coagulation, flake formation is faster due to the high molecular weight of this element, compared to the aluminum; therefore, flakes are denser, and the sedimentation time is significantly reduced (Pavanelli, 2002).

The following reaction, proposed by Pavanelli (2002) refers to the hydrolysis reaction of ferric chloride, being responsible for the formation of iron hydroxide which has a coagulant action on the particles.

$$FeCl_3 + 3H_2O \rightarrow Fe(OH)_3 + 3HCl$$

The use of $FeCl_3$ reduces color, turbidity, amount of suspended solids, BOD drastically and in addition eliminates phosphates.

Delgado et al. (2003) used ferric chloride as coagulant in the treatment of effluent refrigerator, achieved efficiencies that oscillated in range from 60 to 75% for turbidity removal, by applying dosages ranging between 5 and 30 mg L^{-1} during the coagulation process.

Silva et al. (2007) studied and compared ferric chloride and moringa in the post-treatment of effluents from anaerobic reactors of sludge blankets, and even concluding that the ferric chloride showed to be more efficient in the removal of chemical oxygen demand (COD), turbidity and also financially more viable. However, it had iron levels above that those established by legislation, being indispensable another treatment for coagulant removal.

Natural coagulants

Natural coagulants may also be named polyelectrolytes for being polymers originating of proteins and polysaccharides (Barros, 2002).

According to Silva (2005) it has been sought in the biodiversity of natural resources a natural coagulant, biodegradable, with low toxicity, simple use, inexpensive and easily obtained.

Mangrich et al. (2013) said that the sludge resulting from the treatment of water or effluents, since it is organic, can be used as prime material for the production of organic fertilizer, with slow and controlled nitrogen release, thus avoiding urea use, for example.

The use of natural coagulants for effluent treatment can benefit to frigorific industry. In consideration that the effluent generation is unavoidable, the water obtained after treatment could be recycled in industry outside areas and the generated sludge will be more easily eliminated due to their biodegradability.

Moringa seed extract, chitosan and tannin, are among the polyelectrolyte which are more commonly used.

Moringa

The *Moringa oleifera Lam.* is a small arboreal species of fast growth with 5 to 12 meters height, from the Indian northwest, cultivated thanks to their food, medicinal, industrial value and water treatment (Bezerra et al., 2004, Schwarz, 2000).

Lenhari & Hussar (2010) analyzed the physical-chemical treatment of effluent from a food industry comparing the moringa seed and a commercial polymer and observed that with increased commercial polymer concentration, reduction of COD decreases, whereas with the addition of natural coagulant, the efficiency increases. Thus they concluded that the replacement generates large both financial and environmental advantages and for being of natural origin does not make non - renewable resources use such as oil, which these polymers are synthesized.

Moringa seeds can also be used as food. Abdulkarima et al. (2005) concluded that the *moringa oleifera* seed has potential to become a new source of oleic oil.

Silva & Matos (2008) evaluated the characteristics of Moringa seeds used for preparing dispersions for water treatment and found that the dispersions containing seed husk are less homogeneous than those shelled, the removal of oleic content of Moringa seeds can contribute to greater removal of turbidity and a granulometric study of dispersions prepared with Moringa shelled seeds showed that the diameter of the particles composing mass without oil was lower, resulting in higher specific surface. For this reason, dispersion had lower values of apparent color and turbidity.

Chitosan

Chitosan was isolated in 1859 by heating the chitin in concentrated potassium hydroxide solution, resulting in a cationic polyelectrolyte obtained from the deacetylation of chitin, which can be obtained from fungi, especially in species of *Mucor genus,* yeast and the exoskeleton of crustaceans, especially shrimp and crabs. The purification process of the chitin consists of the removal of minerals through acid treatment and alkaline deproteinization and deodorizing with sodium hypochlorite. The production of chitosan can be obtained by partial or complete hydrolysis of the acetyl grouping with concentrated sodium hydroxide solutions or enzymatic hydrolysis, as different methods result in chitosans with different degrees of deacetylation and molecular mass by determining their applicability (Capelete, 2011, Muzzarelli, 1986, Tolaimatea et al., 2003, Weska et al., 2007, Wibowo et al., 2007;).

Santos et al. (2003) mentions various applications of chitosan as water treatment, production of cosmetics, medications, food additives, semipermeable membranes and biomaterials development.

Due to its biodegradability chitosan appears as a potential polymer for replacement of synthetic materials widely used in effluents treatment, attempting to the recycle of these wastes (Wibowo et al., 2007).

For Renault et al. (2009) compared to metal salts, chitosan is more efficient at lower concentrations, producing larger flakes, thereby facilitating the sedimentation speed, amount of sludge volume produced is lower and cause less environmental impact due to their biodegradability however this efficiency it is limited in a range of pH and concentration.

Divakaran & Pillai (2001) obtained results where chitosan promoted the maximum clarification of suspensions containing kaolinite (clay) in water, solutions with turbidity ranging from 10 to 160 NTU and efficiency in coagulation was with turbidity fixed in 40 NTU and pH ranging between 6.5 and 7.5.

Gonçalves et al. (2008) used chitosan for the treatment of effluents contaminated with food coloring, and found that the pH reduction from 7 to 6 and increased concentration of chitosan from 250 to 500 mg L^{-1} leads to an increase in the removal of coloring from 33% to over 90%.

Laus et al. (2006) studied acidity removal, iron and manganese from water contaminated by coal mining using chitosan microspheres crosslinked with tripolyphosphate. The results were satisfactory, because the microspheres assisted in remediation of acidity (pH from 2.5 to 6.0), which showed promising to be effective in removing iron and manganese from water, with 100% and 90% removal, respectively.

Tannin

Tannin is a natural coagulant extracted from plant bark as Acácia Negra, for example, which is cultivated in Brazil, only in the state of Rio Grande do Sul. The tannin extraction made for coagulant production to water treatment, uses from 20 to 30% of the bark. This exhausted shell is partly intended for composting to organic fertilizer production. The rest is used in the own factory for char, steam production and electric power generation, in many cases, sufficient to provide all working industry (Mangrich et al., 2013).

Tannin acts in colloidal systems neutralizing charges and forming bridges between these particles, this process is responsible for flake formation and subsequent sedimentation (Martinez, 1996). Barradas

(2004) reports that among their properties, tannin does not change the pH of treated water since it does not consume the environment alkalinity, while it is effective in a wide pH range, from 4.5 to 8.0.

Cruz et al. (2005) stated that the use of a renewable prime material such as vegetable tannins, have a lower contribution of sulfate anions to the final sludge, lowest sludge mass generation, and obtaining an organic sludge with greater ease of elimination.

Coral et al. (2009) analyzed aluminum sulfate and tannin of Tanfloc®, brand using concentrations of 10 to 60 $mg.L^{-1}$ and obtained similar results in both coagulants, but with the advantage of the latest does not have remaining metal in the treated water and sludge generated at the end of the process.

According to Sánchez-Martín (2010) in studies conducted in a pilot plant of surface water treatment, the use of Tanfloc® showed color reduction of about 50%, removing surfactants up to 75%, and removal of organic matter represented by the reduction of 40% in COD, and 60% in BOD.

Pelegrino (2011) studied the use of tannin in a system of post treatment of sewage effluent with a 65 mg L^{-1} concentration of tannin and 2.0 mg L^{-1} of cationic polymer, obtained for the parameters studied, satisfactory results with a reduction of 95.2% of turbidity, apparent color with a decreased 82.1%, total phosphorus with 49.2%, COD 80.7% and reduction of 87.9% for total suspended solids.

Cruz (2004) evaluated the use in tannin for the treatment of effluents from an industrial laundry and obtained the best result with the 166 mg L^{-1} concentration of tannin in the pH range between 5.5 and 7.5 obtaining removal more than 80% of organic load and surfactants. After treatment, it was even reviewed, the effluent not only conform the current legislation as presented slightly toxic to *Daphnia similis* (microcrustacean).

Bongiovani et al. (2010) conducted a study where was evaluated the benefits of obtaining drinking water using the Tanfloc® SS and concluded that it is an alternative technique, highlighting the benefits to public health by not showing traces of metals and environmental preservation, because the sludge is biodegradable facilitating their disposal in soil.

Chemicals coagulants x Natural coagulants

In coagulation methods for water and wastewater treatment, conventionally, are used inorganic coagulants from chemical origin, consisted of iron salts and aluminum such as aluminum sulfate ($Al_2(SO_4)_3$), ferric sulfate ($Fe_2(SO_4)_3$) and ferric chloride ($FeCl_3$) (Coral et al., 2009).

Pavanelli & Bernard (2002) studied the chemical coagulants performance in high turbidity water and low true color and aluminum chloride obtained most efficiently in a widely pH range and ferric sulfate results in a lower cost to achieve the same remaining turbidity.

Matos et al. (2007) studied the treatment of washing wastewater and stripping / pulping coffee fruits using four different coagulants, which three were from inorganic origin (aluminum sulfate, chlorinated ferrous sulfate and ferric chloride) and one from organic origin (moringa seed extract). The concentrations used for inorganic were from 0 to 3.0 g L^{-1} and 0 to 60 mg L^{-1} for moringa seed extract and pH ranging from 4.0 to 8.0. The results indicated great potential for the use of moringa seed extract showing greater removal of suspended solids in the pH range from 4.0 to 5.0 and 10 mL L^{-1} concentrations.

The difference between metal coagulant and cationic polymer is in its hydrolytic reaction with water. In polyelectrolytes, polymerized chains are already formed when they are added in the liquid medium. In the metal coagulants, polymerization is initiated on the contact with the liquid medium, followed by colloid adsorption step that exist in the medium (Philippi, 2001).

Cardoso et al. (2008) mentioned that Asian, African and South American countries are using various plants as coagulants/flocculants for the purpose of obtaining drinking water, thus minimizing the environmental and public health problems caused by coagulant from chemical origin.

The use of natural products in the coagulation process it is economically viable, because besides provides efficient removal of various pollutants there is the possibility of the residual sludge reuse, helping to minimize environmental impacts, the costs of acquisition of chemicals, which are imported in some cases (Belisário et al., 2009).

The use of renewable prime materials such as vegetable tannins, have a lower contribution of anions sulfates to the final sludge, lower sludge mass generation, and obtaining an organic sludge with greater ease of elimination (Cruz et al., 2005).

Conclusion

Natural coagulants are increasingly being used for the treatment of frigorific effluents making the treated effluent contains no metal traces, deriving from treatments with chemical coagulants. Among the coagulants above, only chitosan was less efficient due to its dilution best occurs at more acidic pHs, thereby the effluent treatment with this coagulant results in a higher cost to the industry.

References

Abdulkarim SM, Long K, Lai OM, Muhammad SKS, Ghazali HM (2005) Some physico-chemical properties of Moringa oleifera seed oil extracted using solvent and aqueous enzymatic methods. Food Chemistry 93(2) 253–263.

Barradas JLD (2004) Tanino - Uma solução ecologicamente correta: agente floculante biodegradável de origem vegetal no tratamento de água. Novo Hamburgo: Publicação Técnica.

Barros MJ, Nozaki J (2002) Redução de poluentes de efluentes das indústrias de papel e celulose pela floculação/coagulação e degradação fotoquímica. Quimíca Nova. 25(5): 736-740.

Belisário M, Borges PS, Galazzi RM, Del Piero PB, Zorzal PB, Ribeiro AVFN, Ribeiro JN (2009) Emprego de resíduos Naturais no tratamento de efluentes contaminados com fármacos poluentes. Inter Science Place. 1(10).

Bezarra AME, Momenté VG, Medeiros Filho S (2004) Germinação de sementes e desenvolvimento de plântulas de Moringa (*Moringa oleifera* Lam.) em função do peso da semente e do tipo de substrato. Horticultura Brasileira. 22(2): 295-299.

Bongiovani MC, Konradr-Moraes LC, Bergamasco R, Lourenço BSS, Tavares CRG (2010) Os benefícios da utilização de coagulantes naturais para a obtenção de água potável. Acta Scientiarum Technology. 32(2): 167-170.

Brazil. CONAMA – Conselho Nacional do Meio Ambiente (2011) Resolução Nº 430, de 13 de maio de 2011. Disponível em: <http://www.mma.gov.br/port/conama/legiabre.cfm?codlegi=646> (Acesso em: 23 nov 2013).

Brazil. MAPA - Ministério da Agricultura, do Abastecimento e da Reforma Agrária (1995) Portaria Nº 711, de 1º de novembro de 1995. Disponível em:
<http://www3.servicos.ms.gov.br/iagro_ged/pdf/714_GED.pdf>
(Acesso em: 23 nov 2013).

Budd GC, Hess AF, Shorney-Darby H, Neemann J, Spencer CM, Bellamy JD, Hargette PH (2004) Coagulation aplications for new tretment goals. Journal of American Water Works.

Capelete BC (2011) Emprego da Quitosana como Coagulante no Tratamento de Água Contendo *Microcystis aeruginosa* – Avaliação de Eficiência e Formaçao de Trihalometanos. Universidade de Brasília - UnB (Dissertação de Mestrado em Tecnologia Ambiental e Recursos Hidricos).

Cardoso KC, Bergamasco R, Sala Cossich E, Kondadt Moraes LC (2008) Otimização dos tempos de mistura e decantação no processo de coagulação/floculação da água bruta por meio da *Moringa oleifera Lam.* Acta Scientiarum. Technology. 30(2): 193-198.

CETESB - Companhia de Tecnologia de Saneamento Ambiental (2008) Guia Técnico Ambiental de Frigoríficos Industrialização de carnes (bovina e suína). São Paulo.

Coral LA, Bergamasco RR, Bassetti FJ. (2010) Estudo da Viabilidade de Utilização do Polímero Natural (TANFLOC) em Substituição ao Sulfato de Alumínio no Tratamento de Águas para Consumo. 2 nd International Workshop, Advances in Cleaner Production.

Cruz JGH (2004) Alternativas para a aplicação de coagulante vegetal à base de tanino no tratamento do efluente de uma lavanderia industrial. Universidade Federal do Rio Grande do Sul – UFRGS (Dissertação de Mestrado Profissionalizante em Engenharia).

Cruz JGH, Menezes JCSS, Rubio J, Schneider IAH (2005) Aplicação de coagulante vegetal à base de tanino no tratamento por coagulação/floculação e adsorção/coagulação/floculação do efluente de uma lavanderia industrial. 23º Congresso Brasileiro de Engenharia Sanitária e Ambiental. Campo Grande – Mato Grosso.

Da Silva MRA, Oliveira MC, Nogueira RFP (2004) Estudo da aplicação do processo foto-fenton solar na degradação de efluentes de indústria de tintas. Eclética Química. 29(2):19-25.

Delgado S, Diaz F, Garcia D, Otero N (2003) Behaviour of inorganic coagulants in secondary effluents from a conventional wastewater treatment plant. Filtration and Separation. 40(7):42-46.

Divakaran R, Pillai VNS (2001) Flocculation of kaolinite suspensions in water by chitosan. Water Research. 35(16): 3904-3908.

Di Bernardo L, Dantas ADB (2005) Métodos e técnicas de tratamento de água. 2ª ed. v. 1. RiMa.

Fernandes MLM (2007) Produção de lípases por fermentação em estado sólido e sua utilização em biocatálise. Universidade Federal do Paraná – UFPR (Tese de Doutorado em Química).

Muzzarelli R, Jeuniaux C, Gooday GW. (1986) Chitin in nature and technology. New York: Plenum Press. p. 129-139.

Gogate PR, Pandit AB (2004) A review of imperative technologies for wastewater treatment II: hybrid methods. Advances in Environmental Research. 8(3-4):553-597.

Gonçalves JO, Vieira MLG, Piccin JS, Pinto LAA (2008) Uso de quitosana no tratamento de águas contaminadas com corante alimentício. XVII Congresso de Iniciação Científica e X Congresso de Pós- Graduação. Universidade Federal de Pelotas - UFL.

Hassemer MAN, Sens ML (2002) Tratamento do efluente de uma indústria têxtil. Processo físico-químico com ozônio e coagulação/floculação. Engenharia Sanitária e Ambiental. 7(2): 30-36.

Hemkemeier M, Betto TL, Manfron R (2009) Caracterização e tratamentos físico-químico de efluente líquido de laboratório de análise de leite. Revista CIATEC – UPF. 1(1): 32-51.

Huang H, Schwab K, Jacangelo JG (2009) Pretreatment for low pressure membranes in water treatment: a review. Environmental Science and Technology. 43(9):3011-3019.

IBGE - Instituto Brasileiro de Geografia e Estatística (2013) Estatística da Produção Pecuária. Disponível em: <http://www.ibge.gov.br/home/estatistica/indicadores/agropecuaria/pr

oducaoagropecuaria/abate-leite-couro-ovos_201303_publ_completa.pdf> (Acesso em: 28 mar 2014).

Jordão EP, Pessoa AC (1995) Tratamento de esgotos domésticos, concepções clássicas de tratamento de esgotos, 2ªedição. CETESB, São Paulo. 544p.

Jung F, Cammarota MC, Freire DMG (2002) Impcat of enzymatic pré-hidrolysis on batch actives sludge systems dealing with oily wastewaters. Biotecnology Letters. 24(21):1797-1802.

Laus R, Laranjeira MCM, Martins AO, Fávere VT, Pedrosa RC, Benassi JC, Geremias R (2006) Microesferas de quitosana reticuladas com tripolifosfato utilizadas para remoção da acidez, ferro (III) e manganês(II) de águas contaminadas pela mineração de carvão. Química Nova. 29(1):34-39.

Leiknes TO (2009) The effect of coupling coagulation and flocculation with membrane filtration in water treatment: a review. Journal of Environmental Sciences. 32(1):8-12.

Lenhari JLB, Hussar GJ (2010) Comparação entre o uso da *Moringa Oleifera Lam* e de polímeros industriais no tratamento fisicoquímico do efluente de indústria alimentícia. Engenharia Ambiental. Espírito Santo do Pinhal. 7(4):033-042.

Macêdo JAB (2001) Subprodutos do processo de desinfecção de água pelos derivados clorados - Disinfection by products – DBP. Macêdo.

Machado BJF (2005) Reuso de efluentes em torres de resfriamento – estudo de caso: Aeroporto Internacional do Rio de Janeiro. Universidade Federal do Rio de Janeiro – UFRJ (Dissertação de Mestrado em Tecnologia de Processos Químicos e Bioquímicos).

Mangrich AS, Doumer ME, Mallmann AS, Wolf CR (2013) Química Verde no Tratamento de Águas: Uso de Coagulante Derivado de Tanino de *Acacia mearnsii*. Revista Virtual de Química. 20(20).

Martinez FL. Taninos vegetais e suas aplicações. Universidade de Havana, Cuba.

Matos AT, Cabanellas CFG, Cecon PR, Brasil MS, Mudado CS (2007) Efeito da Concentração de coagulantes e do pH da solução na turbidez da água, em recirculação, utilizada no processamento dos frutos do cefeeiro. Engenharia Agrícola, Jaboticabal. 27(2): 544-551.

Moraes LCK (2004) Estudo da coagulação-ultrafiltração com o biopolímero quitosana para a produção de água potável. Universidade Estadual de Maringá – UEM (Dissertação de Mestrado em Engenharia Química).

Naime R, Garcia AC (2005). Utilização de Enraizadas no Tratamento de Efluentes Agroindustriais. Estudos tecnológicos. 1(2):9-20.

Pavanelli G, Bernardo L (2002) Eficiência de diferentes tipos de coagulantes na coagulação, floculação e sedimentação de água com turbidez elevada e cor verdadeira baixa. VI Simpósio Ítalo Brasileiro de Engenharia Sanitária e Ambiental.

Pelegrino ECF (2011) Emprego de coagulante à base de tanino em sistema de pós-tratamento de efluente de reator UASB por flotação.

Universidade de São Paulo – USP (Dissertação de Mestrado em Hidráulica e Saneamento).

Phillipi AJ (2001) Desenvolvimento de um equipamento para testes de floculação através de floculação em meio granular expandido. Universidade Federal de Santa Catarina – UFSC (Dissertação de Mestrado em Engenharia Ambiental).

Renault F, Sancey B, Dabot PM, Crini G (2009) Chitosan for coagulation/flocculatio processes – an eco-friendky approach. European Polymer Journal. 45(5):1337-1348.

Sánches-Martín J, Beltrán-Heredia J, Solera-Hernández C (2010) Surface water and wastewater treatment using a new tannin-based coagulant. Pilot plant trials. Journal of Environmental Management. 91(10):2051–2058.

Santos JE, Soares JP, Dockal ER, Campana Filho SP, Cavalheiro ÉTG (2003) Caracterização de quitosanas comerciais de diferentes origens. Revista Polímeros: ciência e tecnologia, São Carlos, SP. 3(4).

Santos EPCC (2004) Coagulação da água da represa vargem das flores visando tratamento por filtração direta. Universidade Federal de Minas Gerais - UFBH (Dissertação de Mestrado em Saneamento, Meio Ambiente e Recursos Hídricos).

Santos EPCC, Teixeira AR, Almeida CP, Libânio M, Pádua VL (2007) Estudo da coagulação aplicada à filtração direta descendente. Engenharia Sanitária Ambiental, Rio de Janeiro. 12(4).

Scarassati D, Carvalho RF, Delgado VL, Coneglian CMR, Brito NN, Tonso S, Sobrinho GD, Pelegrini R (2003) Tratamento de efluentes de matadouros e frigoríficos. III Fórum de Estudos Contábeis. Centro Superior de Educação Tecnológica (CESET) – UNICAMP.

Schneider RP, Tsutiya MT (2001) Membranas Filtrantes para o Tratamento de Água, Esgoto e Águas de Reuso. ABES. 234p.

Schoenhals M, Sena FR, José JH (2006) Avaliação da eficiência do processo de coagulação/floculação aplicado como tratamento primário de efluentes de abatedouros de frangos. Engenharia Ambiental. Espírito Santo do Pinhal. 3(2):005-024.

Schwarz D (2000) Water Clarification using Moringa oleifera. Disponível em: < http://stipulae.johnvanhulst.com/DOCS/PDF/Gate_Moringa.pdf> (Acesso em: 17 mar 2014).

Silva CA (2005) Estudos aplicados ao uso da *moringa oleifera* como coagulante natural para melhoria da qualidade de águas. Universidade Federal de Uberlândia – UFU (Dissertação de Mestrado em Química).

Silva FJA, Matos JEX (2008) Sobre dispersões de Moringa oleifera para tratamento de água. Revista Tecnologia Fortaleza. 29(2):157-163.

Silva MER, Aquino MD, Santos AB (2007) Pós-tratamento de efluentes provenientes de reatores anaeróbios tratando esgotos sanitários por coagulantes naturais e não-naturais. Revista Tecnologia Fortaleza. 28(2):178-190.

Skoog DA, West DM, Holler FJ (1994) Analytical chemistry. 6. ed. Saunders College Pub.

Tolaimatea A, Desbrieresb J, Rhazia M, Alaguic A (2003) Contribution to the preparation of chitins and chitosans with controlled physico-chemical properties. Polymer. 44(26):7939-7952.

Vidal G, Carvalho A, Méndez R, Lema JM (2000) Influence of the content in fats and proteins on the anaerobic biodegradability of dairy wastewater. Bioresource Technology. 74(3):231-239.

Weska RF, Moura JM, Batista LM, Rizzi J, Pinto LAA (2007) Optimization of deacetylation in the production of chitosan from shrimp wastes - Use of response surface methodology. Journal of Food Engineering. 80(3):749-753.

Wibowo A, Velazquez G, Savant V, Torres JA (2007) Effect of chitosan type on proteis and water recovery efficiency from surimi wash water treated with chitosan-alginate complexes. Bioresource Technology. 98(3):539-545.

I want morebooks!

Buy your books fast and straightforward online - at one of the world's fastest growing online book stores! Environmentally sound due to Print-on-Demand technologies.

Buy your books online at
www.get-morebooks.com

Kaufen Sie Ihre Bücher schnell und unkompliziert online – auf einer der am schnellsten wachsenden Buchhandelsplattformen weltweit! Dank Print-On-Demand umwelt- und ressourcenschonend produziert.

Bücher schneller online kaufen
www.morebooks.de

SIA OmniScriptum Publishing
Brivibas gatve 1 97
LV-103 9 Riga, Latvia
Telefax: +371 68620455

info@omniscriptum.com
www.omniscriptum.com

Printed by Books on Demand GmbH, Norderstedt / Germany